Disclaimer

The publisher of this book is by no way associated with the National Institute of Standards and Technology (NIST). The NIST did not publish this book. It was published by 50 page publications under the public domain license.

50 Page Publications.

Book Title: Development of Soft Armor Conditioning Protocols for NIJ 0101.06: Analytical Results

Book Author: Amanda L. Forster; Kirk D. Rice; Michael A. Riley; Guillaume Messin; Sylvain H. Petit; Cyril Clerici; Gale A. Holmes; Joannie W. Chin

Book Abstract: This publication details the four major phases of analytical development work, coupled with several additional side studies, undertaken by the Office of Law Enforcement Standards in writing the Flexible Armor Conditioning Protocol in NIJ 0101.06. This protocol partially fulfills a requirement to develop a revised performance standard for body armor to address a number of concerns, one of which was the ability of the armor to withstand environmental and wear conditions that an armor might see over its lifetime. All major classes of materials were tested without incident, and the conditions selected are found to be quite detrimental to armors of a design that previously had problems in the field. However, the protocol does not represent an exact period of time in the field.

Citation: NIST Interagency/Internal Report (NISTIR) - 7627

Keyword: weapons and protective systems; artificial aging; ballistic fiber; body armor; service life prediction; polymers; polymeric material degradation

NISTIR 7627

Development of Soft Armor Conditioning Protocols for NIJ Standard–0101.06: Analytical Results

Amanda L. Forster, Kirk D. Rice, Michael A. Riley,
Guillaume H. R. Messin, Sylvain Petit, Cyril Clerici, Marie-Cecile
Vigoroux, Pierre Pintus, Gale Holmes, and Joannie Chin

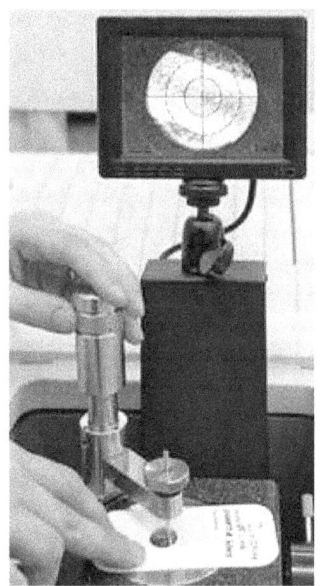

National Institute of
Standards and Technology
U.S. Department of Commerce

NISTIR 7627

Development of Soft Armor Conditioning Protocols for NIJ Standard–0101.06: Analytical Results

Amanda L. Forster, Guillaume H. R. Messin, Pierre Pintus,
Michael A. Riley, and Kirk D. Rice
Office of Law Enforcement Standards
Electronics and Electrical Engineering Laboratory

Gale A. Holmes
Polymers Division
Materials Science and Engineering Laboratory

Joannie Chin, Sylvain Petit, Cyril Clerici, Marie-Cecile Vigoroux
Materials and Construction Research Division
Building and Fire Research Laboratory

September 24, 2009

U.S. Department of Commerce
Gary Locke, Secretary

National Institute of Standards and Technology
Patrick D. Gallagher, Director

Abstract

This publication details the four major phases of analytical development work, coupled with several additional side studies, undertaken by the Office of Law Enforcement Standards in writing the Flexible Armor Conditioning Protocol in NIJ Standard–0101.06. This protocol partially fulfills a requirement to develop a revised performance standard for body armor to address a number of concerns, one of which was the ability of the armor to withstand environmental and wear conditions that an armor might see over its lifetime. This document details how the protocol was shortened from 9 weeks in the first phase of development to 10 days as it appears in the current version of the standard. All major classes of ballistic materials were tested in the protocol development. The conditions selected are found to be quite detrimental to armors of a design that previously had problems in the field, but are not detrimental to armors of known good design. It is important to note that the protocol does not represent an exact period of time in the field, but efforts to correlate the protocol to a period of time in the field are the subject of future research.

This page intentionally left blank.

Acknowledgments

Financial support for this research effort was provided by the National Institute of Justice under Interagency Agreement Number 2003-IJ-R-029. Their support is gratefully acknowledged.

This page intentionally left blank.

Disclaimer

Certain commercial equipment, instruments, or materials are identified in this paper in order to specify the experimental procedure adequately. Such identification is not intended to imply recommendation or endorsement by the National Institute of Standards and Technology, nor is it intended to imply that the materials or equipment identified are necessarily the best available for this purpose.

This page intentionally left blank.

Contents

1 Introduction — 1
  1.1 Background — 2
  1.2 Theoretical Approach — 3
    1.2.1 Definition of Wear Environment — 3
    1.2.2 Selection of Temperature — 5
    1.2.3 Selection of Relative Humidity — 6
    1.2.4 Simulation of Mechanical Wear — 8

2 Experimental — 11
  2.1 Extracted Yarn Tensile Testing — 11
  2.2 Fourier Transform Infrared Analysis — 12
  2.3 Moisture Sorption Analysis — 12
  2.4 Dynamic Mechanical Thermal Analysis — 12

3 Conditioning Protocol Development — 15
  3.1 Phase I — 15
    3.1.1 Sample Description — 15
    3.1.2 Experimental Conditions — 16
    3.1.3 Analytical Results — 17
  3.2 Phase I Summary — 23
  3.3 Phase II — 24
    3.3.1 Sample Description — 24
    3.3.2 Experimental Conditions — 24
    3.3.3 Analytical Results — 25
    3.3.4 Ballistic Results from Phase I and Phase II Testing — 26
  3.4 Phase II Summary — 26
  3.5 Phase III — 26
    3.5.1 Experimental Conditions — 30
    3.5.2 Sample Description — 30
    3.5.3 Analytical Results — 31
    3.5.4 Ballistic Results from Phase III Testing — 36
  3.6 Phase III Summary — 36
  3.7 Phase IV — 36
    3.7.1 Sample Description — 36
    3.7.2 Analytical Results — 37
  3.8 Important Observations From Other Studies — 38
  3.9 Phase IV Summary — 39

4 Conclusions and Future Work	41
5 References	43

List of Tables

1.1 "Rule of Thumb" for Kinetics of Chemical Reactions. . . . 4

3.1 Rank Ordering of PPTA Infrared Bands; 1=greatest change; 4=least change.. 23

This page intentionally left blank.

List of Figures

1.1 Dynamic Temperature Ramp Results for Common Ballistic Fibers. 5
1.2 Representative Vehicle Temperature and Relative Humidity Data. 6
1.3 Moisture Sorption Data for PPTA (cyclic conditions). . . . 7
1.4 Moisture Sorption Data for PBO (cyclic conditions). . . . 8
1.5 Moisture Sorption Data for PBO (constant conditions). . . 9
1.6 Moisture Sorption Data for PPTA (constant conditions). . . 10
3.1 Phase I Protocol Cycle. 16
3.2 Phase I Breaking Strength Retention for PBO. 18
3.3 Phase I Breaking Strength Retention for PPTA. 19
3.4 Phase I FTIR Difference Spectra for PBO. 20
3.5 Phase I FTIR Difference Spectra for PPTA (full scale). . . 21
3.6 Phase I FTIR Difference Spectra for PPTA (expanded scale). 22
3.7 Phase II Protocol Cycle. 25
3.8 Phase II Breaking Strength Retention for PPTA and PBO (all conditions). 27
3.9 Phase II Breaking Strength Retention for PPTA and PBO (T & RH only). 28
3.10 Phase II Breaking Strength Retention for PPTA and PBO (tumbling only). 29
3.11 Phase III Breaking Strength Retention for PPTA and PBO (all conditions). 32
3.12 Phase III Breaking Strength Retention for PPTA and PBO (T & RH only). 33
3.13 Phase III Breaking Strength Retention for PPTA and PBO (tumbling only). 34
3.14 Phase III Reduction in Key Infrared Bands. 35
3.15 Phase IV Breaking Strength Retention for PPTA and PBO (all conditions). 37
3.16 Analysis of Potential for Condensation at 70 °C and 90% Relative Humidity. 39
3.17 Analysis of Potential for Condensation at 65 °C and 80% Relative Humidity. 40

This page intentionally left blank.

1 Introduction

In response to the 2003 US Attorney General's initiative to examine failures of soft body armor containing the material poly p-phenylene-2, 6-benzobisoxazole, or PBO, the National Institute of Justice (NIJ) determined that a significant revision of the performance standard for ballistic body armor was required. One area that had not previously been examined was the long term, or field performance of body armor. Historically the National Institute of Standards and Technology's (NIST) Office of Law Enforcement Standards (OLES) has been NIJ's technical partner in the development of performance standards for body armor. OLES and NIST had also worked closely with NIJ to examine the issues with PBO in the field and published several papers [1,2,3] and reports documenting the degradation of PBO fiber with exposure to elevated conditions of moisture and temperature. Once the issues with PBO fibers became clear, NIJ issued "NIJ Body Armor Standard Advisory Notice # 01-2005" to inform the community of body armor end users about the degradation issues with PBO. Concurrently, NIJ issued the "NIJ 2005 Interim Requirements for Bullet-Resistant Body Armor," requiring manufacturers to state that their armor did not contain any material listed on an NIJ Standard Advisory Notice (e.g. PBO), and requiring that the armor "will maintain ballistic performance (consistent with its originally declared threat level) over its declared warranty period." Subsequently NIJ turned to OLES to develop a revised performance standard for body armor to address a number of concerns, one of which was the ability of the armor to withstand environmental and wear conditions that an armor might see over its lifetime. This document describes the development of a soft armor conditioning protocol to address this requirement.

Previous work published at NIST documented a detailed examination of the failure of an officer's PBO armor in the field [4]. Two key observations from this study were that yarns extracted from the officer's armor showed a 32 % reduction in tensile strength when compared with yarns extracted from new armor, and that infrared spectroscopy analysis of yarns from the officer's vest showed evidence of degradation in the molecular structure of PBO. Further studies at NIST examined degradation of PBO armors under controlled laboratory conditions. A crucial finding from these studies was that PBO fibers degrade when exposed to elevated moisture and temperature but are stable when exposed to elevated temperature in a dry environment [5]. Studies [6,7] also showed that PBO yarns were vulnerable to degradation by mechanical wear, showing classic fatigue behavior. Findings from all of this fundamental research formed the basis of the theory behind the soft armor conditioning protocol.

The primary goals of the soft armor conditioning protocol for use in NIJ Standard–0101.06 are to develop a test protocol that would have caught the problems with PBO-based soft body armor before they appeared in the field and to ensure that the protocol will neither under- nor over- expose armor with respect to the environment that armor is expected to encounter during its lifetime. It quickly became clear that relating this protocol to an exact period of time in the field would be impossible. Body armor is made up of many different materials, all of which show different rates of degradation with exposure to a given set of conditions. To date, very little work has been published on artificial accelerated aging of fibers used in body armor. Work is currently underway at NIST to develop the relationship between exposure at conditions of low temperature to conditions at high temperature. Currently, including the aspect of mechanical wear in this relationship still remains a challenge. Because of this ongoing work and the challenges involved in developing correlations between field and laboratory aging, this protocol will not predict the service life of body armor. Currently, the soft armor conditioning protocol in NIJ Standard–0101.06 can be considered a "challenge test" for armor that provides an indication as to whether or not the armor will withstand use in the field. This represents a major change from previous versions of the ballistic body armor standard, which gave no consideration to long term performance.

1.1 Background

Historically, there have been several efforts to assign an expected service life to body armor. Two studies are typically cited, one undertaken by DuPont in the mid-1980s [8] and one undertaken by NIST (then the National Bureau of Standards, NBS) published in 1986 [9]. The DuPont study indicated that a reduction in ballistic performance as measured by ballistic limit, V_{50} testing, was seen after 3 to 5 years of use, but that a reduction in performance was better correlated to heavy use than to the age of the poly(phenylene terephthalamide) or PPTA, armor. As a result of this study, DuPont recommended that armor be replaced after 5 years, which caused some controversy in the law enforcement community [8, 10, 11, 12, 13, 14]. The NBS study examined 24 sets of 10 year old armor of the same 100 % woven PPTA design, manufactured at the same time and distributed to various law enforcement agencies. The sample set of armors was distributed across various climates and saw various levels of wear, encompassing a range from never issued to heavily worn. The author concluded that armor stored under warehouse conditions maintained its full ballistic performance for at least 10 years and perhaps indefinitely. The author also concluded that light to moderate wear may improve ballistic performance, and that heavy wear might slightly reduce ballistic performance. It is important to note that the limited sample size of this study makes it difficult to draw meaningful conclusions about the long term performance of armor in the field [9].

In more recent years, several body armor manufacturers have undertaken programs to examine the performance of fielded PBO armor by retrieving vests from the field, assigning a wear rating to the vests, and then conducting ballistic limit testing on the vests. Two reports one from Armor Holdings Product Division [15] and one from DHB Armor Group [16] were both published in 2004. Both reports concluded that there was some loss in ballistic performance with both age and wear of the armor, although the methods used to report the data make it difficult to draw meaningful conclusions about the results. Both armor manufacturers indicated that they felt that used armor still had an adequate margin of safety. A study was also undertaken between 2001 and 2005 by the Technical Support Working Group (TSWG) to examine the effect of environmental conditions on armor performance by exposing shoot packs of various ballistic materials to elevated conditions of moisture and temperature. TSWG operates as a program element under the Department of Defense Combating Terrorism Technical Support Office (CTTSO) and they serve as "the national interagency research and development program for combating terrorism requirements at home and abroad." Unfortunately, due to problems with controlling the exposure conditions used in this study, the results were inconclusive. After reviewing the limited body of work that had been conducted on armor service life prediction, it was determined that there was little available to draw on for the development of the soft armor conditioning protocol.

1.2 Theoretical Approach

1.2.1 Definition of Wear Environment

In an effort to better tailor a revision of NIJ Standard–0101.04 to the needs of the end user community, NIJ issued a Request for Information (RFI) to the armor community, including manufacturers and end users, in the fall of 2005. The RFI stated that "... NIJ is interested, though not exclusively, in operational requirements and testing methodologies that address: Validation of used armor performance; Non-destructive testing/monitoring methods for used armor to ensure ongoing performance; Improved requirements and testing protocols for new armor (e.g., blunt trauma, multi-shot impacts, contact shots); Numbers and sizes of samples to be tested; Long-term performance of armor; Artificial armor aging protocols to replicate field use; Quality control and conformity assessments. . ." In reviewing the responses to this request, several respondents were contacted. One of these, Mine Safety Appliance (MSA), had a long history in the production and service life prediction of other types of safety equipment. In January 2006, a meeting between MSA and NIST was held to discuss armor aging, in which several approaches were discussed. In this discussion, it was suggested that armor should be robust enough to withstand conditions typically seen during wear and those seen during transit. Based on the previously published guidelines for armor replacement, a typical service life was defined as 5 years,

Temperature		Exposure	
°C	°F	h	weeks
35	95	10,000	59.5
45	113	5000	29.8
55	131	2500	14.8
65	149	1250	7.44
75	167	625	3.72

Table 1.1: "Rule of Thumb" for Kinetics of Chemical Reactions.

and a typical wear environment was defined as near body temperature and humidities near complete saturation (due to perspiration of the wearer). one defines a typical work schedule as 8 h per day, 5 days per week, 50 weeks per year, this works out to 2000 h of wear per year. If one then expects the typical lifetime of a vest to be 5 years, then that corresponds to 10,000 h of service at the wear conditions. These simple assumptions provided a starting point for the development of the protocol. To maintain the independence of NIST and NIJ, no further input was sought from MSA after these initial meetings in the protocol development [17].

A "rule of thumb" in chemical kinetics [18] often applied to accelerated aging of materials is for every 10°C increase in temperature, one can expect a doubling in the rate of reaction. Application of this guideline to the defined wear temperature of 35 °C, results in 10,000 h of aging in approximately 8 weeks at 65 °C, as shown in Table 1.1. It is important to note that this "rule of thumb" applies to certain reactions that occur in solution and does not directly translate to reactions of degradation in the solid state. Additionally, body armor is made up of many different types of materials, all of which can be assumed to degrade at different rates. So, while the temperature 65°C was chosen to accelerate degradation in the armor based on assumptions of a 5 year service life, it definitely cannot be said to predict armor service life because we do not know the exact relationship between temperature and degradation rate for ballistic materials.

When attempting to apply the "Rule of Thumb" to accelerated aging kinetics of materials, it is important to keep in mind that elevated temperatures may induce new mechanisms of degradation, rather than accelerating mechanisms of lower-temperature degradation. For example, if temperature is increased to the point that a material would melt or burn, different chemical reactions will occur, and the results of the accelerated study will not be meaningful. To avoid this potential problem, dynamic mechanical thermal analysis (DMTA) was performed on representative fiber samples from the major material classes of body armor prior to temperature selection. A dynamic temperature ramp at a constant frequency of 1 Hz was performed on fiber samples of PPTA, PBO, and ultrahigh molecular weight polyethylene (UHMWPE). These results are presented in Figure 1. This analysis revealed that temperatures exceeding 80°C might be too high due to changes

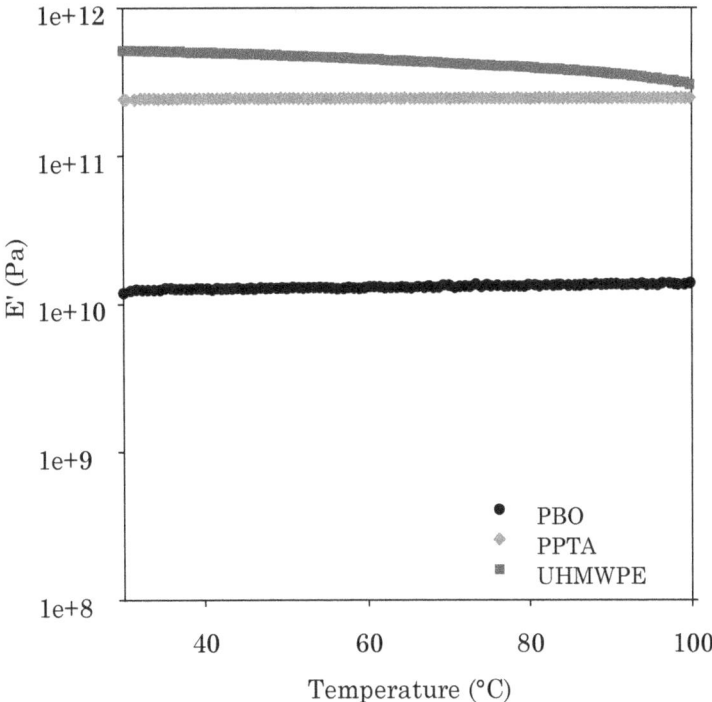

Figure 1.1: Dynamic Temperature Ramp Results for Common Ballistic Fibers.

in the molecular structure of the UHMWPE fibers above this temperature, however the PPTA and PBO fibers remain essentially unchanged in the temperature range studied. In the UHMWPE system, 80 °C is in the range of the α'-relaxation temperature, which is the temperature at which molecular motion within the polymer begins to increase, resulting in a decrease in the modulus of the polymer [19]. DSM Dyneema, a manufacturer of UHMWPE fiber for ballistic applications, published results of an artificial aging study in 2007 indicating that an Arrhenius relationship existed for UHMWPE fibers between 35 °C (the same as our reference base temperature) and 65 °C [20]. Therefore, it was determined that limiting our experiments to temperatures below 70 °C would allow us to avoid introducing new mechanisms of degradation in the fibers during our studies.

1.2.2 Selection of Temperature

Questions also arose regarding the exposure of armor to a temperature of 65 °C—(e.g., is this condition unreasonable in the environment in which body armor will be used?). Anecdotal evidence that officers commonly store armor in the trunks of their cars was frequently brought up during discussions of the armor wear environment. Additionally, armor is typically delivered across the country in trucks. However, the temperature and relative humidity inside a vehicle can vary widely depending on season, geographical region of the country, and location inside the vehicle. In order

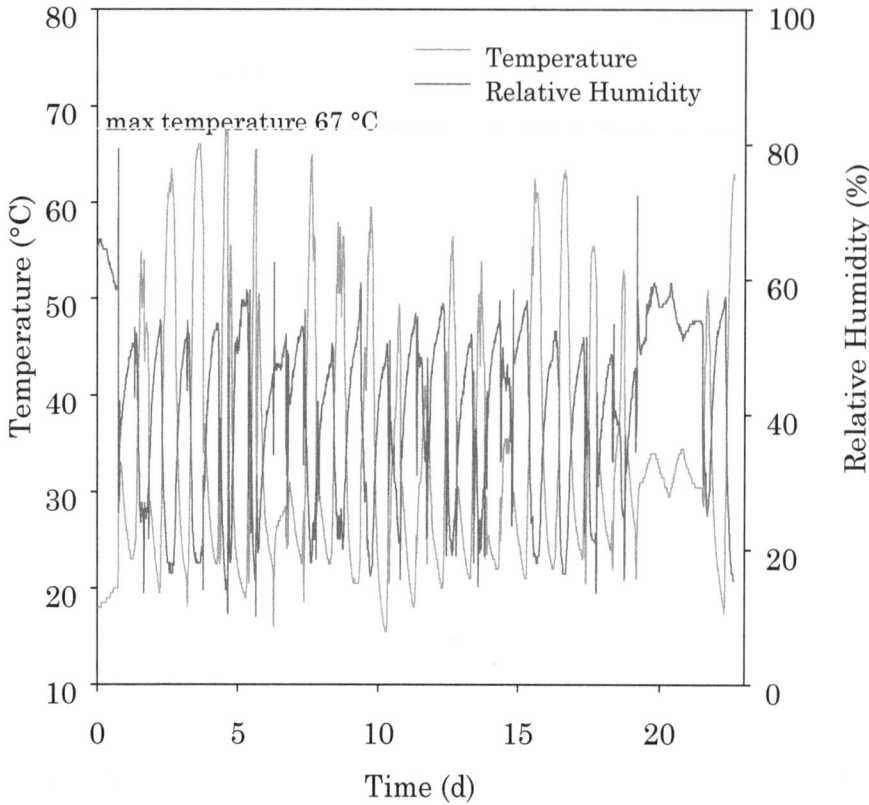

Figure 1.2: Representative Vehicle Temperature and Relative Humidity Data.

to answer this question, cooperation was sought from NIJ's Body Armor Technology Working Group (TWG), which is made up of law enforcement and corrections officers who have interest or expertise in ballistics and body armor. Small, inexpensive universal serial bus (USB)-readable temperature and relative humidity data loggers were purchased and distributed to volunteers from the TWG from across the United States. These were placed inside actual police vehicles throughout different seasons and the data were examined periodically. The same data recorders were also placed inside OLES staff members' personal vehicles during the summer of 2006 in Maryland. A high temperature of 67 °C was obtained in July 2006 from the study of OLES staff member vehicles. Readings around 63 °C were also obtained in California and Illinois during the summer of 2007. An example of representative vehicle data is shown in Figure 1.2.

1.2.3 Selection of Relative Humidity

Another parameter that must be selected is the relative humidity used in the exposure conditions. One of the participants in the TWG vehicle conditions study, independent of the NIST study, obtained permission to have officers wear an environmental sensor on the outside of their armor. Relative

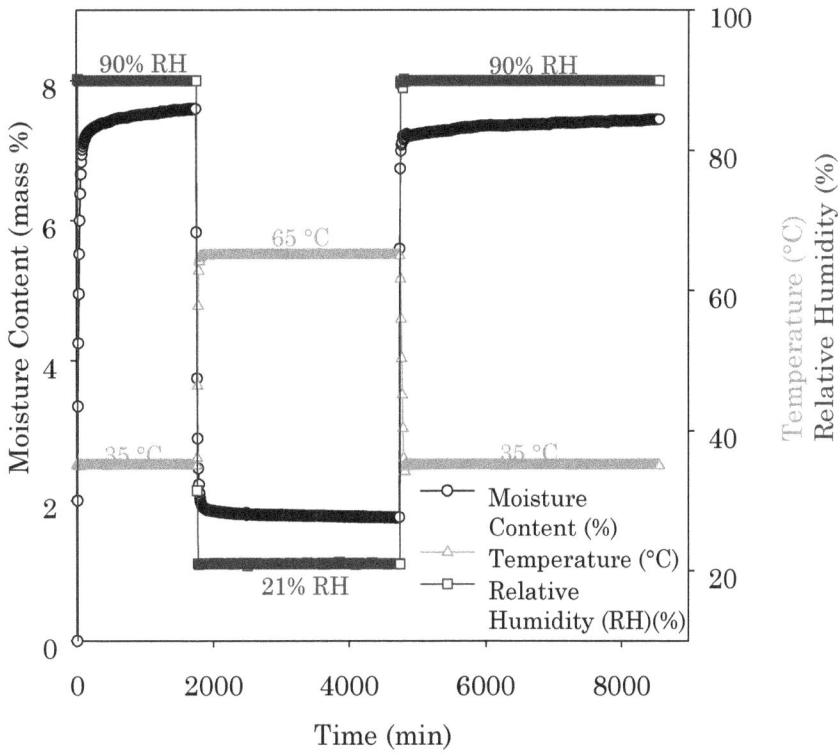

Figure 1.3: Moisture Sorption Data for PPTA (cyclic conditions).

humidities and temperatures outside of the armor are probably close to, but possibly slightly lower than, those seen within the armor. The maximum temperatures seen during this study was 41 °C and the maximum relative humidity seen was 76%. Initially, the protocol was envisoned as a cyclical temperature and relative humidity exposure, with a low temperature condition of 35 °C, 90% relative humidity. In an effort to maintain consistent conditions at an elevated temperature of 65 °C, high temperature relative humidities were envisoned as 21%, which corresponds to the same quantity of water per gram of dry air (0.032 gram water per gram dry air). However, prior to beginning testing, a moisture sorption study was undertaken to examine the moisture uptake by the fiber at the two conditions of temperature and relative humidity. In changing conditions from 35 °C, 90% relative humidity to 65 °C, 21% relative humidity, a large desorption was observed in both PPTA (Fig. 1.3) and PBO (Fig. 1.4) fibers. This indicated that the moisture content in the air around the fiber was actually the wrong variable to control—it was more important to attempt to maintain a constant moisture content in the fiber, where the degradation reactions would occur. After examining several conditions, it was determined that 75% relative humidity at both 35 °C and 65 °C would result in an approximately equal moisture content, as observed for PBO (Fig. 1.5) and PPTA (Fig. 1.6) fibers, so this relative humidity was selected for the initial trials.

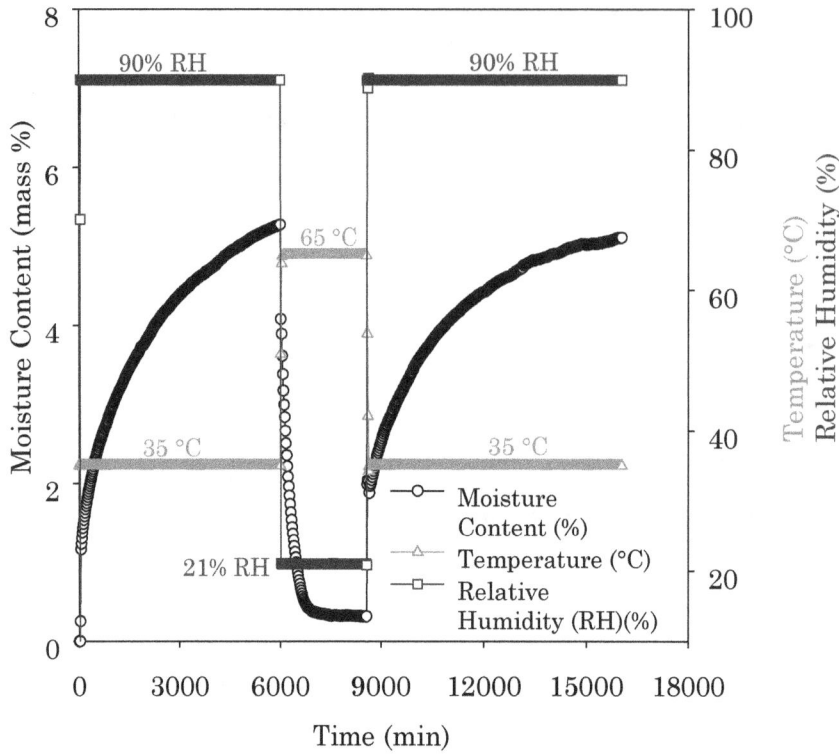

Figure 1.4: Moisture Sorption Data for PBO (cyclic conditions).

1.2.4 Simulation of Mechanical Wear

Determination of the temperature and relative humidity conditions for the soft armor conditioning protocol was relatively straightforward. Defining the wear environment to simulate involved an analysis of possible conditions. However, environmental exposure only provides part of the solution. In the course of normal wear, armor is exposed to flexing, bending, and abrasion. All of these conditions could potentially cause degradation in the ballistic performance of armor. The combination of mechanical conditions with environmental exposure is the overall goal of the soft armor conditioning protocol. However, the definition of the mechanical wear environment is extremely challenging—tests which provide only abrasion ignore the potential fatigue aspects of folding and bending of the armor. A conservative analysis that a body armor user might bend over (e.g., when entering or exiting a vehicle) 4 times per hour, 40 h per week, 50 weeks per year could result in 8000 folding cycles per year, or 40,000 folding cycles over 5 years. Realistically, almost any movement a wearer makes results in some type of bend or fold in the armor, which could add up to many thousands of cycles per year [6]. Significant work has been devoted to this area by Holmes and co-workers at NIST [7]. Tests which create a single fold or bend in the armor create challenges in assessing the ballistic performance of armor, because very little area is available over which to conduct ballistic testing. This would require additional samples and drive up the cost of testing, and

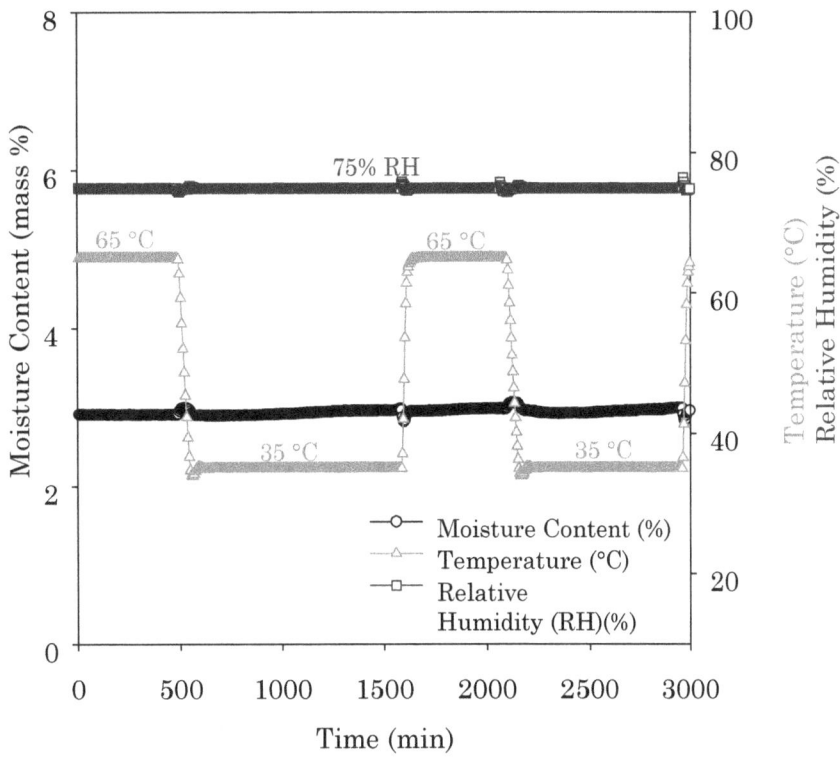

Figure 1.5: Moisture Sorption Data for PBO (constant conditions).

ignores the problem of abrasion, which is more difficult to quantify. The best solution to this challenge is to find a method of creating mechanical wear that roughly simulates the same types of wear seen in the field and provides a relatively uniform level of mechanical wear to the entire armor. Tumbling was selected as the solution that best combined simulating the desired damage with cost efficiency, both in terms of capital cost of equipment and quantities of samples. The goal of combining the tumbling with the environmental exposure was initially challenging, so initial trials were done by removing armor panels from an environmental chamber, and tumbling periodically during the exposure period.

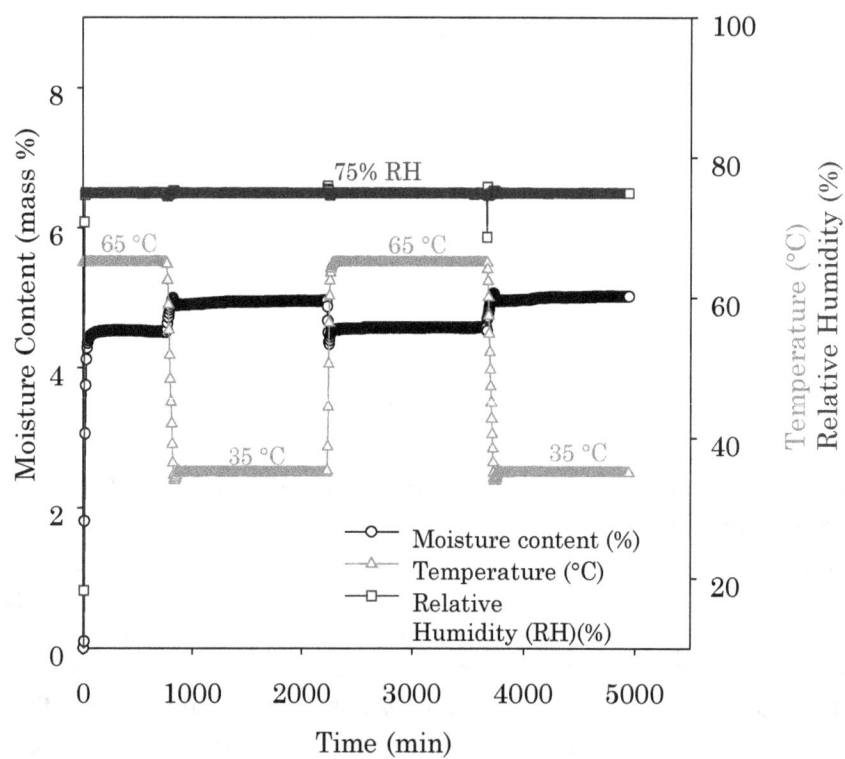

Figure 1.6: Moisture Sorption Data for PPTA (constant conditions).

2 Experimental

A combination of analytical and ballistic characterization techniques were used throughout the development of the soft armor conditioning protocol in an effort to learn as much as possible from each trial. The two analytical techniques that were most commonly used were tensile testing and Fourier transform infrared analysis (FTIR) of yarns extracted from test armor panels. Other analytical techniques that were used included dynamic mechanical thermal analysis (DMTA) and moisture sorption analysis (MSA). Ballistic testing was limited to two techniques—perforation/backface signature (P-BFS) testing and ballistic limit (V_{50}) analysis.

2.1 Extracted Yarn Tensile Testing

To obtain yarn mechanical properties, tensile testing of yarns was carried out in accordance with ASTM D2256-02, "Standard Test Method for Tensile Properties of Yarn by the Single-Strand Method," using an Instron Model 4482 test frame equipped with a 91 kg (200 lb) load cell, and pneumatic yarn and cord grips (Instron model 2714-006). The jaw separation was 7.9 cm (3.1 in) and the cross-head speed was 2.3 cm/min (0.9 in/min). In this study, yarns were nominally 38.1 cm (15 in) long, and given 60 twists[1] on a custom-designed yarn twisting device. This level of twist was maintained on the yarns as they were inserted into the pneumatic yarn and chord grips. Strain measurements were made with an Instron non-contacting Type 3 video extensometer in conjunction with black foam markers placed approximately 2.5 cm apart in the gage section of the yarn. Ten to twelve replicates from each vest were tested to failure. The standard uncertainty of these measurements is typically 3%, however the error bars generated for plots presented herein are based on the relative standard deviation of the yarn breaking strength, which is in some cases higher than 3%.

---

[1] This twist level is within the range recommended by ASTM D2256-02, and was experimentally verified prior to beginning experiments.

2.2 Fourier Transform Infrared Analysis

Infrared analysis was carried out using a Nicolet Nexus Fourier Transform Infrared (FTIR) Spectrometer equipped with a mercury-cadmium-telluride (MCT) detector and a SensIR Durascope attenuated total reflectance (ATR) accessory or a Bruker Vertex 80 FTIR, also equipped with a Smiths Detection Durascope ATR accessory. Air dried by passage through a standard FTIR purge gas generator was used as the purge gas. Consistent pressure on the yarns was applied using the force monitor on the Durascope. FTIR spectra were recorded at a resolution of 4 cm^{-1} between 4000 cm^{-1} and 700 cm^{-1} and averaged over 128 scans. Three different locations on each yarn were analyzed. Spectral analysis, including spectral subtraction, was carried out using a custom software program developed in the Building and Fire Research Laboratory's Polymeric Materials Group at NIST. All spectra were baseline corrected and normalized using the aromatic C-H deformation peak at 848 cm^{-1} for PBO and 820 cm^{-1} for Kevlar. Standard uncertainties associated with this measurement are typically 4 cm^{-1} in wavenumber and 1 % in peak intensity.

2.3 Moisture Sorption Analysis

Moisture absorbed by the yarn specimens during the temperature/humidity exposure period was measured using a Hiden IGAsorp Moisture Sorption Analyzer. The IGAsorp software monitors the temporal changes in the mass of a specimen subjected to prescribed temperature and relative humidity conditions, and calculates equilibrium parameters via curve fitting. Specimens for sorption analysis were prepared by disassembling between 5 mg and 7 mg of yarn into individual filaments to prevent capillarity effects from dominating the sorption process. Prior to beginning a sorption experiment, specimens were dried in the moisture sorption analyzer at $\approx$ 0 % relative humidity and the prescribed temperature at which the sorption experiment would be carried out. Moisture uptake was measured at 50 °C and 60 % relative humidity as well as at 60 °C and 37 % relative humidity. The water sorption isotherm was generated using the isothermal mapper mode at 40 °C within a range of 0 % relative humidity to 95 % relative humidity. Results are the average of two specimens. The standard uncertainty of these measurements is typically 0.02 % mass fraction.

2.4 Dynamic Mechanical Thermal Analysis

Dynamic Mechanical Thermal Analysis (DMTA) was performed using a TA Instruments RSAIII DMTA. Dynamic temperature ramp measurements were generated by loading a single fiber into film/fiber tension clamps, and applying a preload of approximately 1 g force to the sample. The measurement was performed in a strain controlled mode with a strain of 0.1 % at a frequency of 1 Hz. The temperature was increased from 30 °C to 110 °C at a ramp rate of 3 °C/min. For the RSAIII, the manufacturer-stated relative

2. Experimental

standard uncertainty [21] in the force measurement is typically ± 0.0002g, and the standard uncertainty in the temperature scale is typically ± 0.5 °C.

This page intentionally left blank.

3 Conditioning Protocol Development

Multiple phases of development were conducted, all utilizing slightly different methods and equipment configurations. Each phase of development will be presented separately in this document to better describe the development of the soft armor conditioning protocol.

3.1 Phase I

Initial experiments were performed using separate tumbling and environmental exposure steps. This allowed for "proof of concept" of tumbling as a mechanism to provide mechanical wear, and also allowed for exploration of potential environmental conditions.

3.1.1 Sample Description

Two types of test armors were used in the first phase of protocol development. One sample armor was constructed of 20 layers of plain woven 500 denier PBO, with 26 yarns per inch in the horizontal direction and 26 yarns per inch in the vertical direction. The layers of fabric were stitched together in two packs of 10 layers each with a 2.54 cm (1 in) diagonal quilt stitch to form the ballistic package. This ballistic package was then encased in a stitched moisture-permeable fabric cover and inserted into a lightweight poly-cotton carrier to form an armor panel. The other sample armor was constructed of 25 layers of plain woven 500 denier PPTA, with 24 yarns in the horizontal direction and 24 yarns in the vertical direction. The layers of fabric were stitched together in one package with a 3.18 cm (1.25 in) diagonal quilt stitch to form the ballistic package. This ballistic package was then encased in a standard water-repellent treated nylon fabric cover and inserted into a medium-weight poly-cotton carrier to form an armor panel[1]. All armors were manufactured specifically for this study. The PBO armor samples were designed to be NIJ Standard–0101.04 Level IIA compliant.

---

[1] The definitions of panel and armor panel used in this document are intended to be consistent with the definitions of these terms as described in Section 3 of NIJ Standard–0101.06 [22].

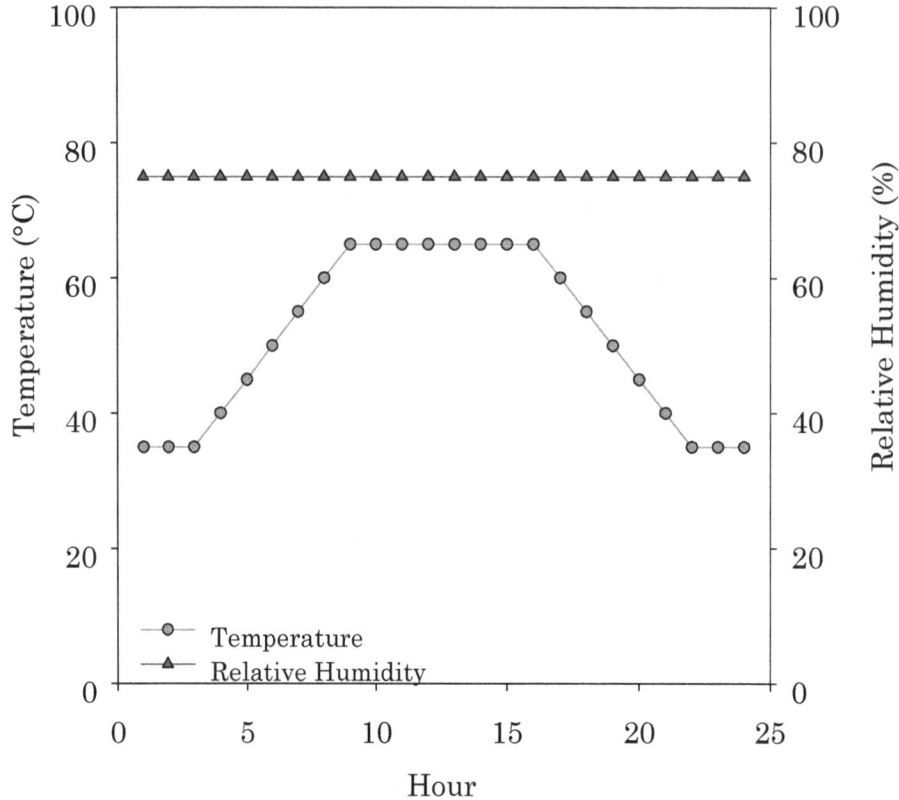

Figure 3.1: Phase I Protocol Cycle.

The PPT armor samples were designed to be NIJ Standard–0101.04 Level II compliant [23]. Both armor samples were constructed to be the size required for NIJ Standard–0101.04 2005 Interim Requirements [24] compliance testing.

Both sample sets consisted of 13 armor panels. Of these 13 panels, 7 panels were exposed to all conditions and were designated for ballistic testing, 1 sample was exposed to all conditions and was used only for analytical testing, 2 panels were controls for heat and moisture and received no tumbling (one of these samples was designated for ballistic testing and one for analytical testing), 2 panels were tumbling controls and received no heat and moisture exposure (1 of these samples was designated for ballistic testing and 1 for analytical testing), and finally 1 control sample received no heat, moisture, or tumbling exposure (was stored at room temperature and humidity of nominally 21 °C and 50% relative humidity) and was used for analytical testing.

3.1.2 Experimental Conditions

As previously discussed, the temperature and relative humidity protocol originally consisted of a cyclical temperature profile between 35 °C and 65 °C, with a constant relative humidity of 75%, as depicted in Figure 3.1.

3. Conditioning Protocol Development

Environmental exposures of the PPTA and PBO samples were conducted in two separate chambers. Vests were hung vertically in the humidity chamber for the environmental portion of the exposures and removed at designated times for tumbling in a standard home clothes dryer (with the heating element disabled). The chamber was returned to room temperature and humidity before removing the armor for tumbling to avoid the formation of a condensing atmosphere in the chamber. Samples were extracted from armor designated for analytical testing after it was removed from the chamber for tumbling. Extractions were performed after the tumbling was completed. An estimate of the total number of revolutions of the armor for the first phase is 194,400 total revolutions, based on 3 h of tumbling, 3 d per week, for 9 weeks. The rotation speed of the standard home clothes dryer was measured at nominally 4.19 rad/s (40 revolutions per minute) using a laser tachometer.

3.1.3 Analytical Results

Relative tensile strengths of yarns extracted from the PBO armor panels are depicted in Figure 3.2. This figure shows the reduction of ultimate tensile strength, plotted as percent strength retention, of the PBO armor as a function of exposure time. After 9 weeks, armor exposed to heat, moisture, and tumbling had a tensile strength retention of approximately 62%. This is comparable to the value that was observed in the back panel of a PBO armor that was defeated in the field, and the value that was ultimately reached after 6 months of aging in a previous study [4]. An interesting observation is that the armor panel that was exposed to only heat and moisture had essentially the same tensile strength retention. The armor panel that was only exposed to tumbling had only a reduction in tensile strength of approximately 8%. This indicates that the tumbling did not produce enough mechanical damage or did not happen frequently enough in the protocol to accelerate the effects of the heat and moisture exposure.

Relative tensile strengths of yarns extracted from the PPTA armor panels are depicted in Figure 3.3. This figure shows the reduction of ultimate tensile strength, plotted as percent strength retention, of the PPTA armor as a function of exposure time. After 9 weeks, the armor exhibits essentially no change in tensile strength and was apparently unaffected by the exposure protocol.

In previous studies [4, 5], it has been shown that oxazole ring opening is a major indicator of hydrolysis in PBO. Oxazole ring opening in previous studies is identified by the loss of peaks attributed to the vibrations associated with the benzoxazole ring at 1496 cm^{-1}, 1362 cm^{-1}, 1056 cm^{-1}, and 914 cm^{-1}, and by the formation of a peak at 1650 cm^{-1} attributed to an amide carbonyl or carboxylic acid, which are potential products of oxazole ring opening. Infrared difference spectra of PBO taken over the course of the exposure study are shown in Figure 3.4. The difference spectra show the chemical changes occurring in PBO as a function of exposure time.

The difference spectra show marked reductions in the peaks at 1492 cm^{-1}, 1361 cm^{-1}, 1056 cm^{-1}, and 914 cm^{-1}, all of which are attributed to oxazole

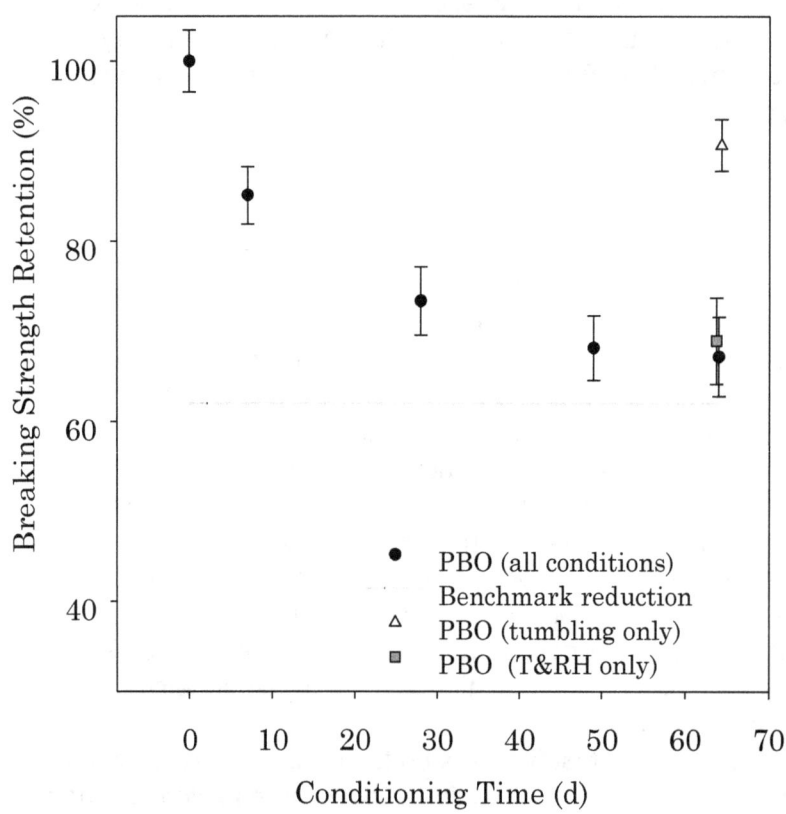

Figure 3.2: Phase I Breaking Strength Retention for PBO. The error bars represent the relative standard deviation of the mean yarn breaking strength. Points are offset horizontally for clarity.

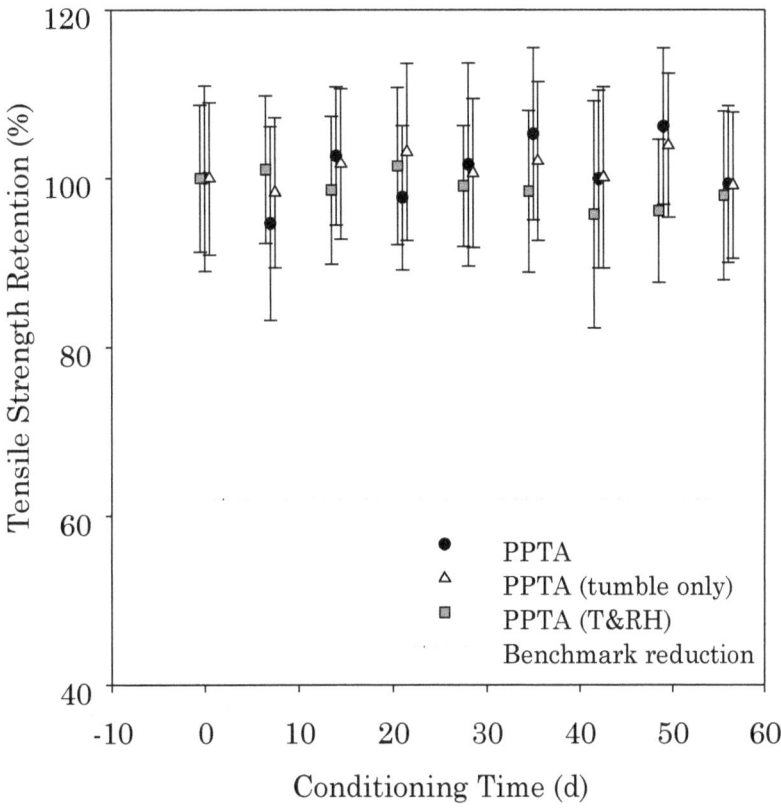

Figure 3.3: Phase I Breaking Strength Retention for PPTA. The error bars represent the relative standard deviation of the mean yarn breaking strength. Points are offset horizontally for clarity.

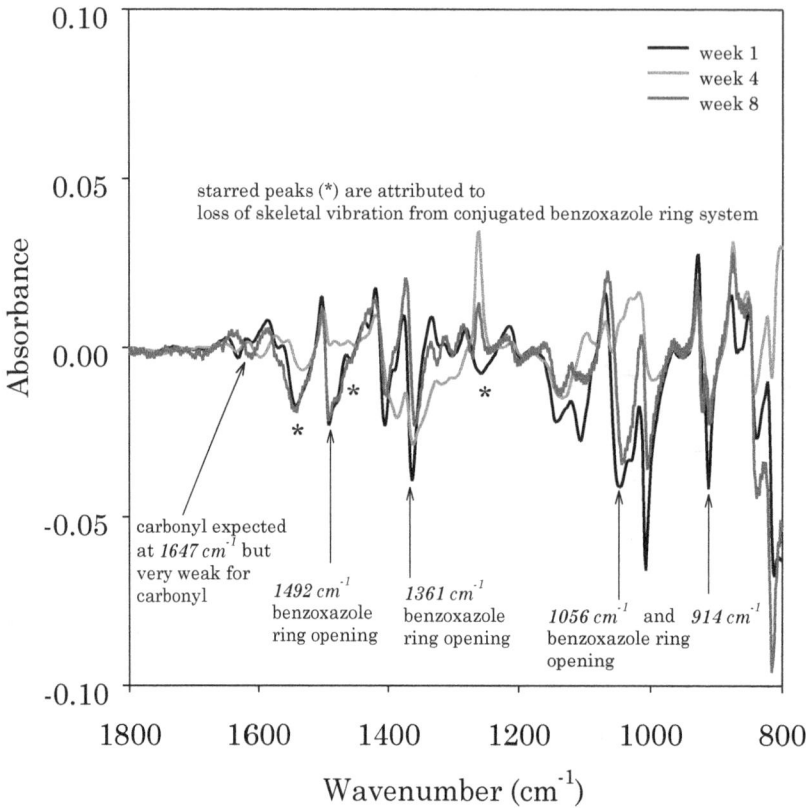

Figure 3.4: Phase I FTIR Difference Spectra for PBO.

ring opening. As previously mentioned, standard uncertainties associated with this measurement are typically 4 cm^{-1} in wavenumber and 1% in peak intensity, so the slight shift in wavenumbers for the difference spectra may be due to variations in the individual spectra used to create the difference spectrum. However, we do not see a large peak at 1650 cm^{-1} to indicate the formation of an amide carbonyl or carboxylic acid as expected. A possible explanation is that this product is leaving the system, possibly due to the abrasion created during tumbling, or that this product was extracted by moisture.

The infrared subtraction spectra of PPTA yarns extracted from body armor panels following environmental exposure are shown in Figure 3.5 and Figure 3.6. The body armor panels were divided into four groups—one group was subjected to tumbling alone, one to temperature/moisture exposure alone, one to temperature/moisture combined with tumbling (designated as "all"), and one group was sealed in plastic bags at room temperature of nominally 22 °C to serve as controls.

Infrared analysis and spectral subtraction revealed that all of the conditions, even the control conditions, resulted in some degree of PPTA hydrolysis. The difference spectra in Figure 3.5 and Figure 3.6 shows negative peaks with positions corresponding to the original amide I peak at 1640 cm^{-1} and amide II peak at 1513 cm^{-1}. A new (positive) broad peak is

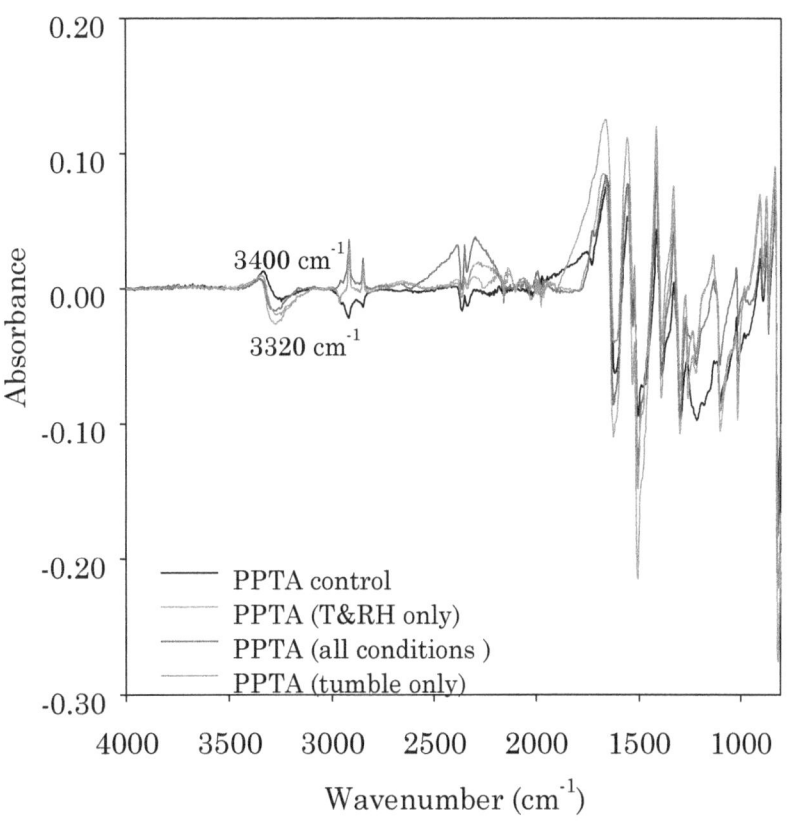

Figure 3.5: Phase I FTIR Difference Spectra for PPTA (fullscale).

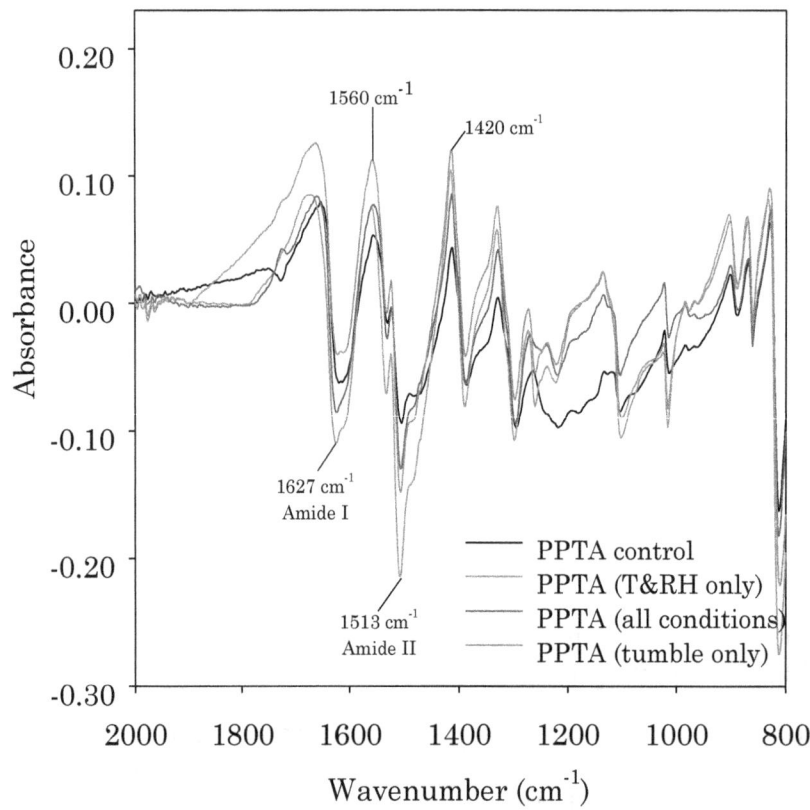

Figure 3.6: Phase I FTIR Difference Spectra for PPTA (expanded scale).

Infrared Band (cm^{-1})	All	T&RH Only	Tumble Only	Control
3320	3	1	2	4
1627	2	1	4	3
1513	3	1	2	4
1560	2(tie)	2(tie)	1	3
1420	2	3(tie)	1	3(tie)

Table 3.1: Rank Ordering of PPTA Infrared Bands; 1=greatest change; 4=least change.

observed at 3400 cm^{-1}, which is attributed to a combination of amine N-H stretching and carboxylic acid OH stretching. New peaks are also observed at 1570 cm^{-1} and 1420 cm^{-1} that are attributed to carboxylate ion stretching. This evidence points to the hydrolysis of the main chain amide group to amine and carboxylic acid.

It is puzzling why even the control panels that were not subjected to any environmental stresses also underwent hydrolysis, albeit to a lesser extent. It is likely that as long as moisture is present, hydrolytic reactions in the PPTAs can occur.

In an attempt to determine which environmental conditioning treatment caused the most hydrolytic damage, intensities of the difference bands were examined and rank-ordered. A tabulation of the major difference bands (except for the bands at 3400 cm^{-1} which were too close to distinguish) and their intensity rankings is given in Table 3.1 below.

No clear or consistent pattern can be found in the above table; the intensity rank order differs for each infrared band. Since the tensile strengths of the yarns extracted from the environmentally conditioned panels did not exhibit any changes over the course of the conditioning treatments, it is possible that these chemical changes are beneath the threshold necessary to influence mechanical properties and may fall within the standard deviations of the infrared measurements.

3.2 Phase I Summary

This phase of development showed that it was possible to develop a protocol of elevated temperature and relative humidity that would cause damage to PBO armor after 9 weeks of exposure, but would not cause similar damage in PPTA armor in this time period. However, the 8 week exposure time was deemed too long for practical implementation in the new NIJ Standard–0101.06. After this first phase, it was determined that the duration of the protocol must be reduced. Reduction of the time required to complete the protocol was the primary goal for Phase II.

3.3 Phase II

As was the case in Phase I, in order to meet the timetable for development of the conditioning protocol, initial exposures were performed using separate tumbling and environmental exposure for Phase II. This allowed for "proof of concept" of tumbling as a mechanism to provide mechanical wear, and also allowed for further exploration of potential environmental conditions.

3.3.1 Sample Description

Three sets of armors were used in the second phase of protocol development. Two woven armors were the same as those discussed in the Phase I testing. The additional armor was a nonwoven armor, constructed of 30 sheets of unidirectional (UD) laminated UHMWPE fibers. In a UD layer, all fibers are laid parallel, in the same plane. In this study, the sheets of UHMWPE armors were made of 4 layers of fibers, with the orientation of fibers in each layer at 90° to the direction of the fibers in the adjacent layers (0°, 90°, 0°, 90°). The sheets of UHMWPE were stitched together in three places at the top of the vest and one place at the bottom of the vest. Phase II also used two chambers and two sample sets. Chamber 1 contained 15 panels of PBO armor and 6 panels of UHMWPE armor. The PBO panels were tested as follows—in chamber 1, one of the PBO samples was a control and was not exposed to any heat, humidity, or tumbling exposure. Three panels were exposed to only tumbling (two of these were for ballistic testing and one was for analytical testing), two panels were exposed to only temperature and relative humidity (one of these was used for ballistic testing and one for analytical testing), 8 panels were exposed to all conditions and were designated for ballistic testing, and 1 sample was exposed to all conditions and was only tested analytically. All six of the UHMWPE panels were tested ballistically. The results of the ballistic testing will be the subject of a separate publication.

Chamber 2 consisted of 15 panels of PPTA armor and 5 panels of UHMWPE armor. The PPTA panels were used as follows—in chamber 2, one of the PPTA samples was a control and was not exposed to any heat, humidity, or tumbling exposure. Three panels were exposed to only tumbling (two of these were for ballistic testing and one was for analytical testing), two panels were exposed to only temperature and relative humidity (one of these was for ballistic testing and one was for analytical testing), 8 panels were exposed to all conditions and were designated for ballistic testing, and 1 sample was exposed to all conditions and was used only for analytical testing. All five of the UHMWPE panels were for ballistic testing.

3.3.2 Experimental Conditions

In an effort to shorten the time required to achieve the target degradation from Phase I, the temperature and relative humidity protocol was adjusted to allow for two cycles within a 24 h time period with a temperature profile between 35 °C and 65 °C, with a constant relative humidity of 75 % except

3. Conditioning Protocol Development

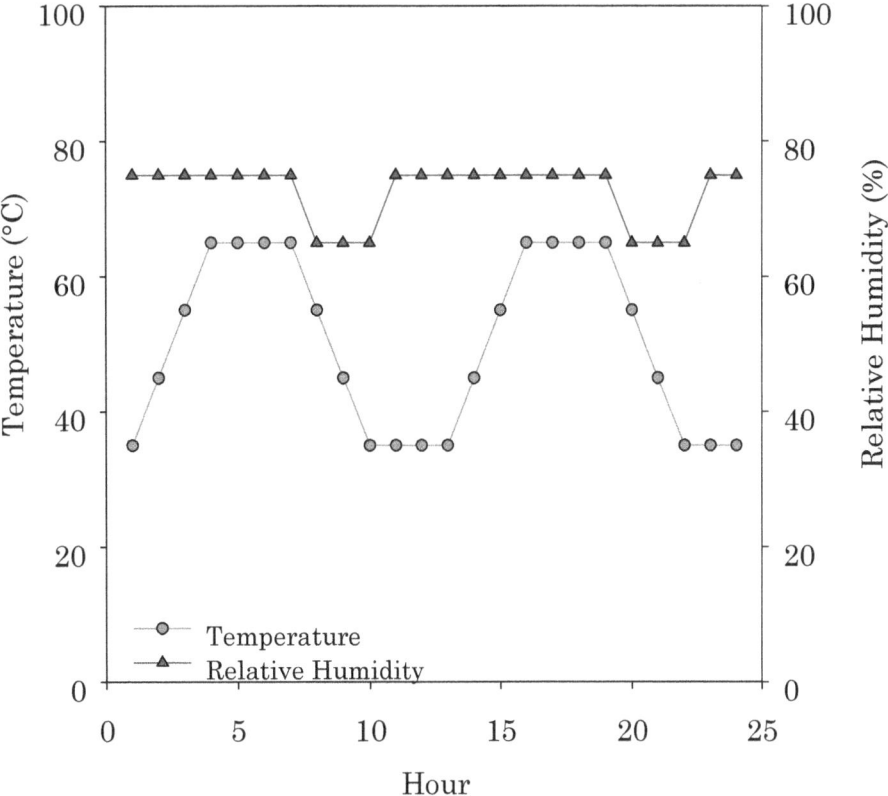

Figure 3.7: Phase II Protocol Cycle.

on the cooling parts of the cycle to avoid condensation, where the relative humidity was dropped to 65% as depicted in Figure 3.7.

Environmental exposures of the PPTA and PBO samples were conducted in two separate chambers. Vests were hung vertically in the humidity chamber for the environmental portion of the exposures and removed periodically for tumbling in a standard home clothes dryer (with the heating element disabled). The chamber was returned to room temperature and humidity before removing armor for tumbling to avoid the formation of a condensing atmosphere in the chamber. Samples were extracted from armor designated for analytical testing when it was removed from the chamber for tumbling. An estimate of the total number of revolutions of the armor for the first phase is 115,200 total revolutions, based on 2 h of tumbling, 2 d per week, for 6 weeks. The rotation speed of the standard home clothes dryer was measured at nominally 4.19 rad/s (40 rpm) using a laser tachometer. The load in each individual tumbler was increased in an effort to increase mechanical damage caused by tumbling.

3.3.3 Analytical Results

Tensile breaking strength testing of yarns extracted from the armor panels are depicted in Figures 3.8 through 3.10. Figure 3.8 shows the reduction of breaking tensile strength, plotted as a percent strength retention, of the PBO armor and PPTA armor as a function of exposure time. After 6 weeks, the PBO sample which was exposed to the conditions of heat, moisture, and tumbling had a tensile strength retention of approximately 58%. This is slightly lower than the target value established by previous studies. By comparison, the PPTA armor showed no reduction in tensile strength in this time. Figure 3.9 shows the results of testing on yarns that had been extracted from vests that were only exposed to temperature and relative humidity. The PBO yarns had an approximate tensile strength retention of 80%, as compared to no strength reduction in the PPTA armor panels. Figure 3.10 shows tensile strength reduction data for armors that were tumbled at room temperature and humidity. Once again, the PBO armor had an approximate tensile strength reduction of 80%, but the PPTA armor was essentially unaffected. It is important to note that for the PBO armors, the panels exposed to only temperature and relative humidity and the panels that were exposed to only tumbling had approximately equal strength retentions. This indicates that the contribution of each mechanism (environmental vs. mechanical) to overall degradation in this study was approximately equal, and that the combination of the two mechanisms had a synergistic effect. Infrared results indicated similar trends to those observed in the previous study and are not included here for brevity.

3.3.4 Ballistic Results from Phase I and Phase II Testing

Ballistic testing, including V_{50} and per fortion/back face signature testing (P-BFS) was performed on armor samples used throughout this study. The results of this testing will be the subject of a separate publication.

3.4 Phase II Summary

In this phase of development, the time per cycle was compressed so that two temperature and humidity cycles were completed per 24 h period, and the armor was tumbled more frequently. However, the protocol was only shortened from 9 weeks to 6 weeks with these changes. The 6 week exposure time was still deemed too long for practical implementation in the new NIJ Standard–0101.06. Therefore, major changes were planned for Phase III in order to further reduce the duration of the test.

3.5 Phase III

Phase III is a very significant phase in the development of Soft Armor Conditioning Protocols because it represents the first phase in which mechanical conditioning was combined with environmental conditioning in one test. Details and specifications of the device which was fabricated in our laboratory at NIST to perform this testing are available via the NIST File Transfer Protocol (FTP) site [25] and will be the subject of a future NIST publication.

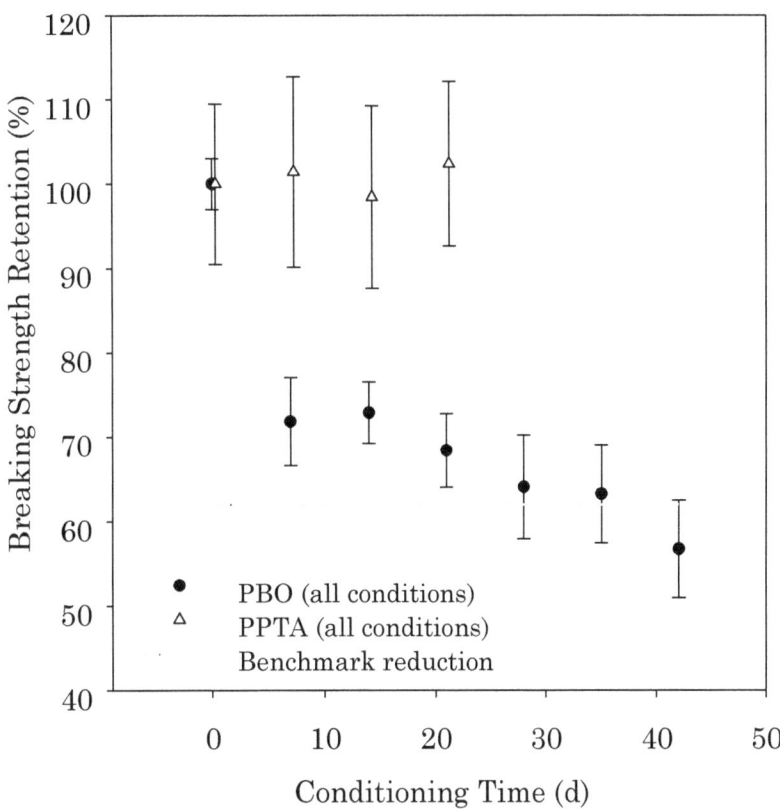

Figure 3.8: Phase II Breaking Strength Retention for PPTA and PBO (all conditions). The error bars represent the relative standard deviation of the mean yarn breaking strength. PPTA points are offset horizontally for clarity.

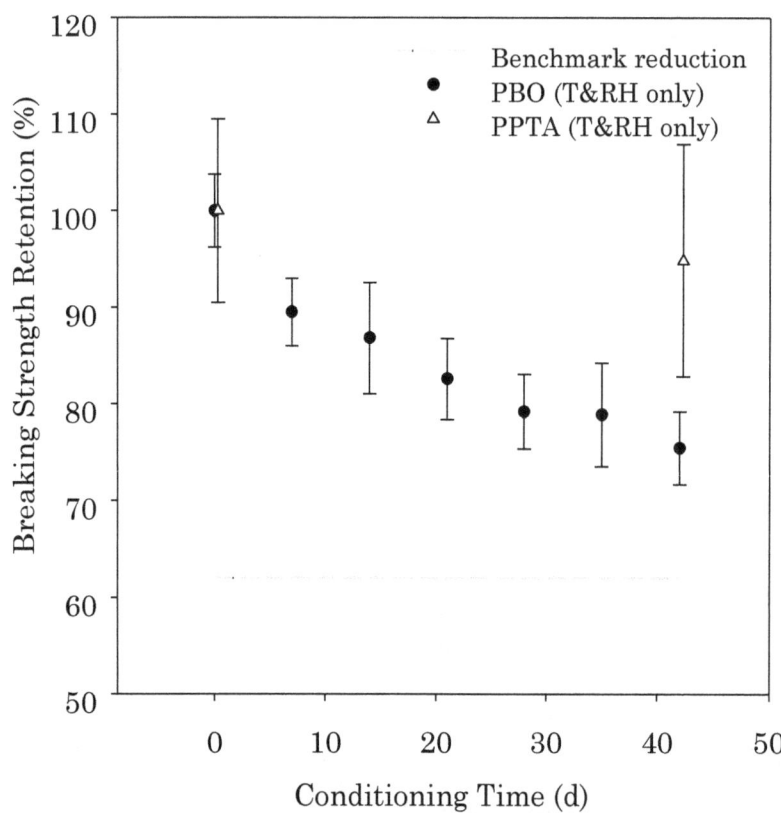

Figure 3.9: Phase II Breaking Strength Retention for PPTA and PBO (T&RH only). The error bars represent the relative standard deviation of the mean yarn breaking strength. PPTA points are offset horizontally for clarity.

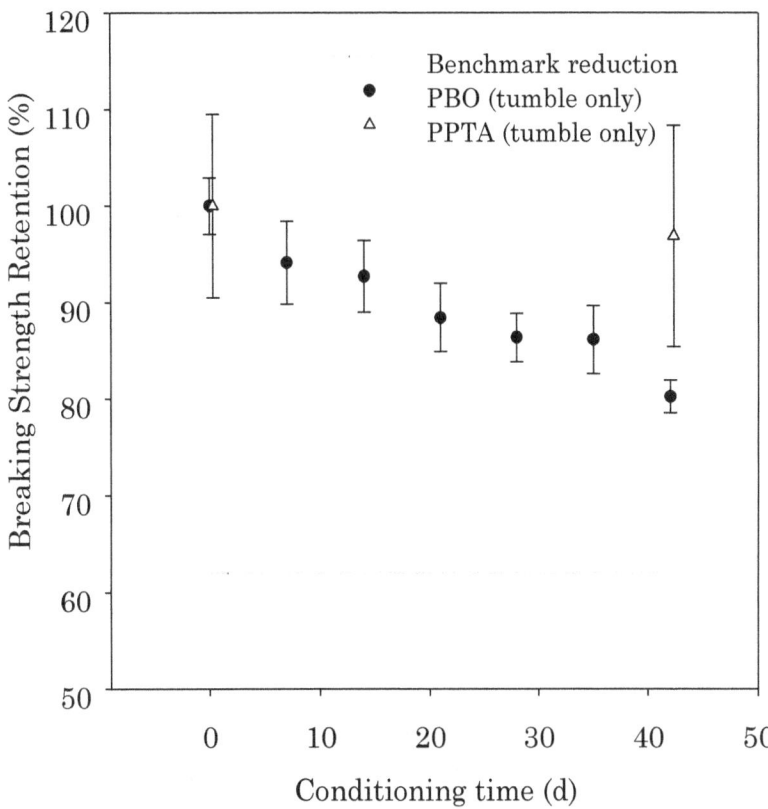

Figure 3.10: Phase II Breaking Strength Retention for PPTA and PBO (tumbling only). The error bars represent the relative standard deviation of the mean yarn breaking strength. PPTA points are offset horizontally for clarity.

3.5.1 Experimental Conditions

In an effort to accelerate the degradation achieved in Phases I and II, the temperature and relative humidity protocol were adjusted to a constant condition of 70 °C and 90% relative humidity. The rationale behind this change was to shorten the test protocol by spending all of the conditioning time at the high heat, high relative humidity condition instead of cycling between two conditions. At this point in the development, it was determined that 8 armors would be tumbled at a time, so efforts were made to try to keep tumbler loads as close to 8 armors, or 16 armor panels, as possible. Tumbling was performed at 0.523 rad/s (5 rpm) continuously throughout the exposure. A specific time interval for exposure was not set at the beginning of the study, the intention was to track the chemical and physical degradation of the armor to determine when the target reduction in tensile strength had been obtained.

3.5.2 Sample Description

Three types of armors were used in the third phase of protocol development. The two woven sample armors were the same as those discussed in the Phase I and Phase II testing. A new set of armor was obtained for Phase III, also constructed of 30 sheets of UD laminated UHMWPE fibers, as described previously. There was no stitching of the sheets of UHMWPE. After ballistic testing was completed, it was discovered that these samples had been fabricated incorrectly. Instead of cutting each sheet of the material separately in order to achieve the correct 0°, 90°, 0°, 90° orientation, the material was rolled out in a back and forth direction and all layers were cut out at once. Additionally, the layers were not aligned properly and portions of layers were missing from all of the armor panels that were manufactured in this way. This construction problem was determined to affect the ballistic properties of the material that were measured after conditioning, as will be discussed in a future publication. One environmental chamber in which a tumbler had been installed was used in Phase III with two sets of samples. The first sample set consisted of 6 PBO armor panels, 7 UHMWPE armor panels, and 6 PPTA armor panels. One of each type of armor panel was exposed to only temperature and relative humidity and used for analytical testing. One of the PBO armor panels and one of the PPTA armor panels that were exposed to all conditions were used for analytical testing. The remaining armor panels (4 PBO, 4 PPTA, and 6 UHMWPE) were designated for ballistic testing. The chamber was programmed at constant conditions of 70 °C and 90% relative humidity, with a constant tumbling speed of 0.52 rad/s (5 rpm). A separate second set of testing was performed to obtain armors that had only been exposed to mechanical conditioning. In this test, 5 PBO panels, 6 UHMWPE panels, and 5 PPTA panels were tumbled at room temperature and humidity at 0.52 rad/s (5 rpm).

3.5.3 Analytical Results

Tensile breaking strength testing of yarns extracted from the PBO and PPTA armor panels are depicted in Figure 3.11 through Figure 3.13. Figure 3.11 shows the reduction of ultimate tensile strength, plotted as a percent strength retention, of the PBO and PPTA armors exposed to all conditions as a function of exposure time. After 10d, the PBO sample that was exposed to the conditions of heat, moisture, and tumbling had a tensile strength retention of approximately 62%, though we continued the test until day 13. In this phase, for the first time, there was an indication of strength loss in the PPTA armor. Samples that were exposed to all conditions had a tensile strength retention of approximately 88% after 13 days of exposure. Figure 3.12 compares the PBO and PPTA armor panels that were exposed to only temperature and relative humidity. The PBO armor panels had a tensile strength retention of approximately 77% and the PPTA armor panels had a tensile strength retention of 90%. Figure 3.13 shows panels that were exposed to only tumbling. The PBO armor panels had a tensile strength retention of approximately 78% and the PPTA armor panels showed no reduction in tensile strength. There are a few conclusions that can be drawn from these results. The first is that for PBO, the test combining environmental exposure and tumbling still had approximately equal contributions of each mechanisms (environmental vs. mechanical) to overall degradation of the material. In the case of the PPTA armor, it is puzzling that there was a slight reduction in tensile strength in the samples that were only exposed to temperature and relative humidity. A possible explanation is that the armor could have been more sensitive to the slightly higher temperature and relative humidity in this study, which could have also been responsible for the slight reduction in strength observed in the armor exposed to all conditions.

Results from infrared analysis of yarns extracted from the PBO armor panels are depicted in Figure 3.14. Similar to the infrared analysis presented in Phases I and II of the study, these results show an overall reduction in the peak absorbance at 1606 cm^{-1}, 1302 cm^{-1}, 1257 cm^{-1}, 1136 cm^{-1}, 1036 cm^{-1}, and 909 cm^{-1}, which are typically associated with the benzoxazole and an increase in the peak absorbance at 1635 cm^{-1}, which is associated with carbonyl formation after opening of the benzoxazole ring. These two trends, taken together, indicate that hydrolysis was achieved in the PBO samples. One may note in Figure 3.14 that there is not a trend indicating a steady decline in absorbance for the benzoxazole ring. There is a sharp decline between day 0 and day 3, then at day 8, there was an apparent increase in the absorbance of the benzoxazole peak. This may be attributed to differences in the samples removed from the armor for testing on the different days, or may be due to the continual mechanical damage occurring in the system due to tumbling. This tumbling may abrade the degraded layer of material and expose fresh material underneath, which would lead to an apparent increase in the absorbance of the benzoxazole peak as referenced to the data obtained on day 3. Analysis of the PPTA samples indicated similar trends to those seen in previous phases of the study.

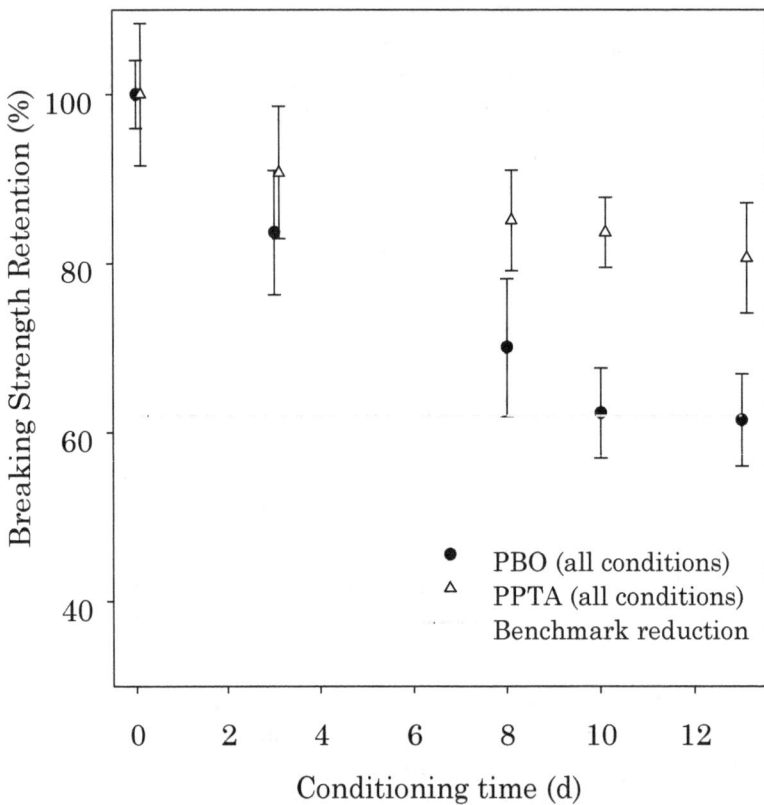

Figure 3.11: Phase III Breaking Strength Retention for PPTA and PBO (all conditions). The error bars represent the relative standard deviation of the mean yarn breaking strength. PPTA points are offset horizontally for clarity.

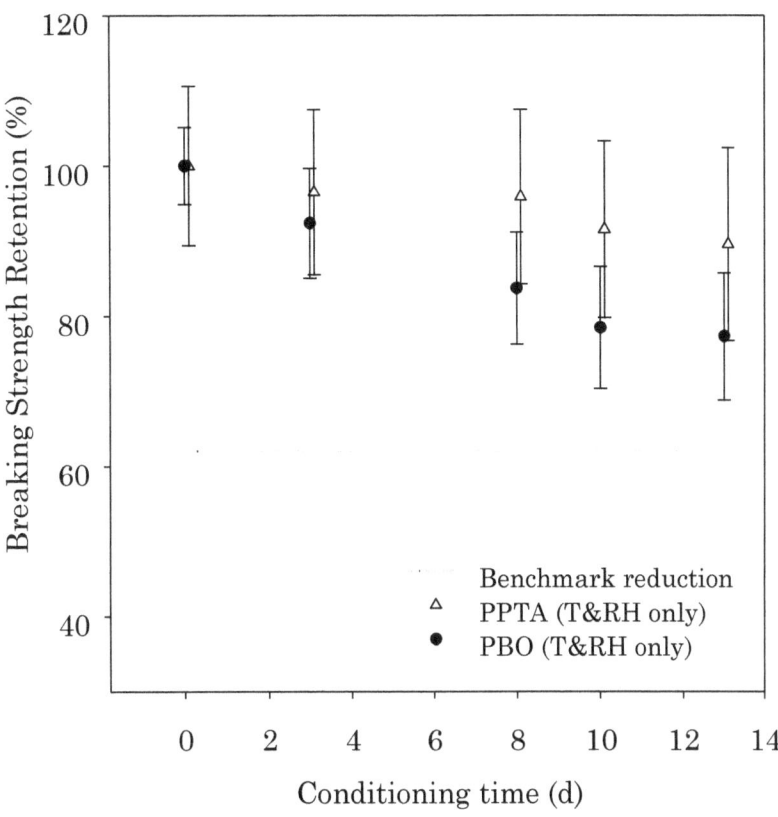

Figure 3.12: Phase III Breaking Strength Retention for PPTA and PBO (T&RH only). The error bars represent the relative standard deviation of the mean yarn breaking strength. PPTA points are offset horizontally for clarity.

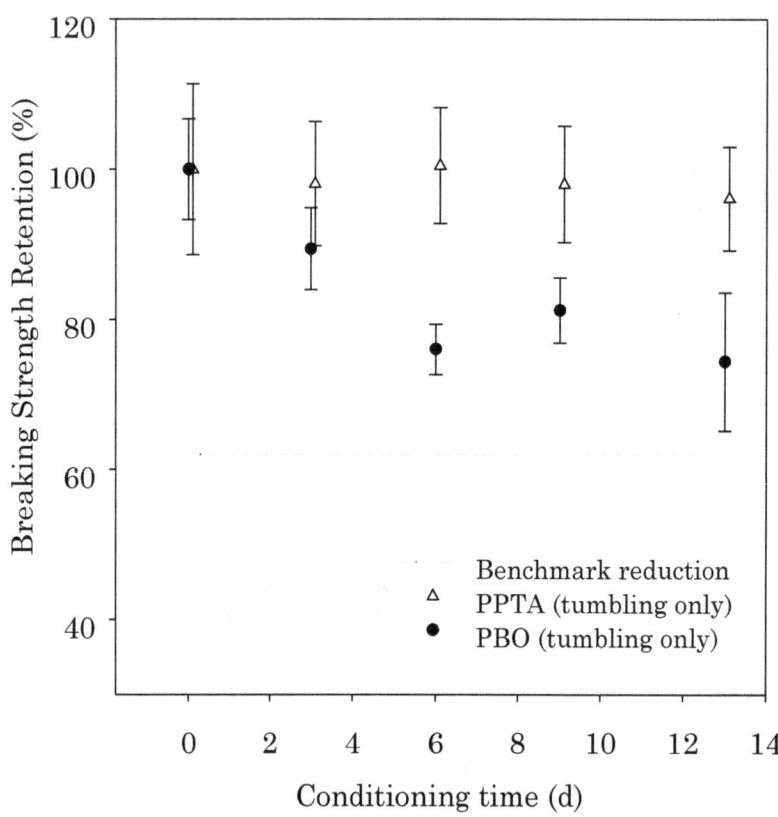

Figure 3.13: Phase III Breaking Strength Retention for PPTA and PBO (tumbling only). The error bars represent the relative standard deviation of the mean yarn breaking strength. PPTA points are offset horizontally for clarity.

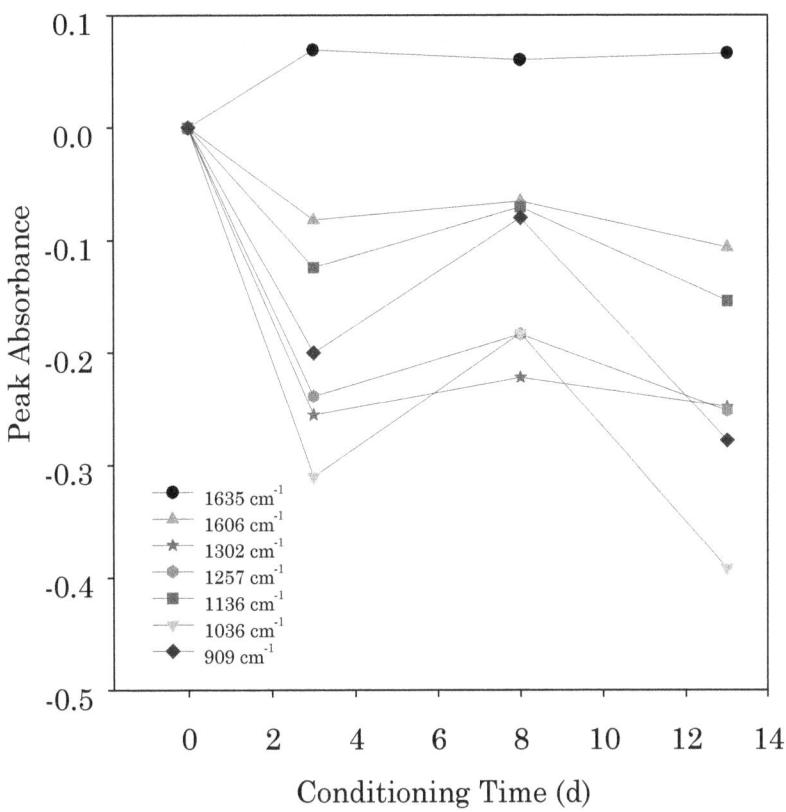

Figure 3.14: Phase III Reduction in Key Infrared Bands.

3.5.4 Ballistic Results from Phase III Testing

Ballistic testing, including V_{50} and perforation/back face signature testing (P-BFS) was performed on armor samples used throughout this study. As previously indicated, the results of this testing will be the subject of a separate publication.

3.6 Phase III Summary

Major changes to the protocol occurred in Phase III. The concept of cyclical conditions of temperature and relative humidity exposure were abandoned in favor of a constant high heat, high relative humidity condition. The tumbling and environmental exposure were combined into one test with the development of a custom-built tumbler inside of the humidity chamber. While the test presented in Phase III ran for 13 days, the target degradation was achieved before the end of the test. Therefore, it was determined that the changes made in Phase III allowed the exposure time to be reduced to a much more practical 10 days. The 10 d test was deemed acceptable for practical implementation into NIJ Standard–0101.06.

3.7 Phase IV

The Phase IV study was designed to verify the conditions selected in Phase III, and verify that 10 d was the appropriate period of time for the test. Conditions of exposure remained the same in this phase as in Phase III.

3.7.1 Sample Description

Three sets of armor samples were used in the fourth phase of protocol development. Two of the armors were the same woven armors as those discussed in the Phase I, Phase II, and Phase III testing. A new set of armor was obtained for Phase IV. This armor model consisted of 18 layers of four plies each of UD PPTA fiber, crossplied at 0°, 90°, 0°, 90° sandwiched between thermoplastic films inside of a nylon armor panel cover. The panel cover seams were heat-sealed and the interior surface of the panel covers were coated for water repellency. There was no stitching of the sheets of UD PPTA. Phase IV used one environmental chamber in which a tumbler had been installed, and one sample set. This sample set consisted of 5 PBO armor panels, 6 UD PPTA armor panels, and 5 woven PPTA armor panels. One of the PBO armor panels and one of the PPTA armor panels which were exposed to all conditions were used for analytical testing. The remaining armor panels (4 PBO, 4 woven PPTA, and 6 UD PPTA) were used for ballistic testing. The chamber was maintained at constant conditions of 70 °C and 90 % relative humidity, with a constant tumbling speed of 0.52 rad/s (5 rpm).

3. Conditioning Protocol Development

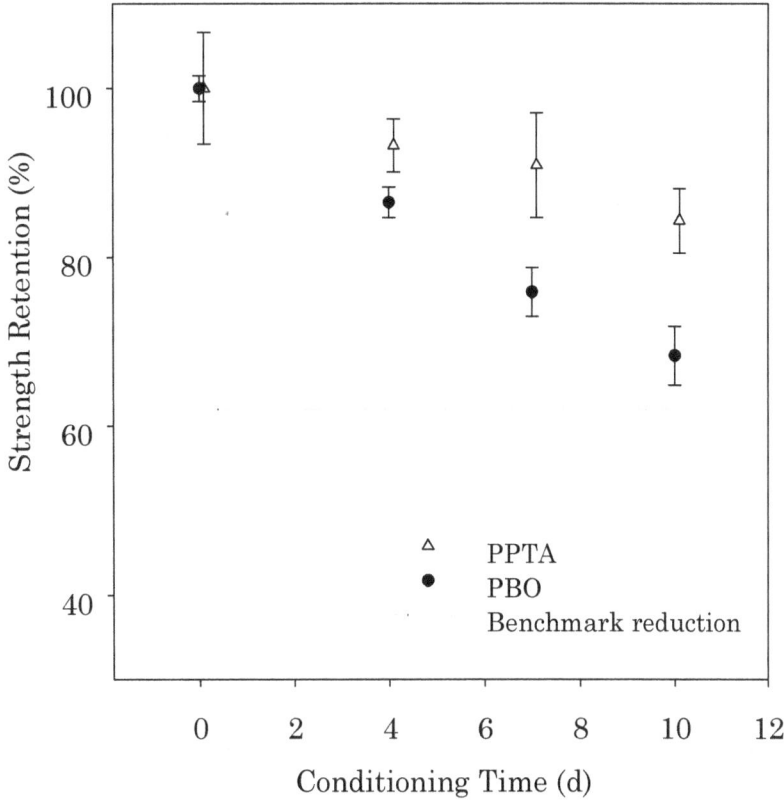

Figure 3.15: Phase IV Breaking Strength Retention for PPTA and PBO (all conditions). The error bars represent the relative standard deviation of the mean yarn breaking strength. PPTA points are offset horizontally for clarity.

3.7.2 Analytical Results

Tensile breaking strengths of yarns extracted from PPTA and PBO armor panels are depicted in Figure 3.15. As discussed in previous trials, this figure shows the reduction of ultimate tensile strength, plotted as a percent strength retention of both types of armor as a function of exposure time. After 10 d, the PBO sample that was exposed to all of the conditions of heat, moisture, and tumbling had a tensile strength retention of approximately 68%. The PPTA sample exposed to all of these conditions had a tensile strength retention of approximately 85%. This strength loss in the PPTA armor was greater than that observed in previous studies. This could potentially be due to variations between the samples tested, or the fact that the operator who performed these tests was not as experienced in performing yarn sample extraction and testing as the previous operators. Infrared analysis of fibers extracted from both the PBO and PPTA armors was performed, but analysis of this data did not provide any additional information beyond what has already been discussed herein.

3.8 Important Observations From Other Studies

Some of the studies performed in development of the Flexible Armor Conditioning Protocol did not result in a full analysis appropriate for treatment within this document. Some of these observations may be important and they will be summarized below. A set of armor was received that had been enclosed in a heat-sealed panel-covering material, but supplied without armor carriers. Some prior work had been performed to evaluate the effect of tumbling with or without a carrier, but only with armors that were encased in sewn panel-covering materials. In these trials, there was no discernible difference between the armors tumbled with and without the carriers. Based on the results of these previous studies, the heat-sealed armors were placed in the tumbler without carriers. However, within a few days, the edges of the heat-sealed panel-covering material started to peel away from the main body of the panel cover. This caused the remaining panel cover to separate and exposed the ballistic package to the tumbler. This experience was one of the factors that led to the decision that all armors would be tested in a specified, generic carrier in NIJ Standard–0101.06.

After the completion of Phase IV testing, a concern was raised that, even within the specified tolerances of relative humidity and temperature, a condensing atmosphere could be achieved within the tumbler. In response to this concern, the saturated vapor curve for water vapor was examined in relation to the specified conditions. Figure 3.16 shows this analysis for conditions of 70 °C and 90% relative humidity, with tolerances of $\pm$ 2 °C and $\pm$ 5% relative humidity. It was discovered from this analysis that if the temperature dropped from the upper limit of 72 °C to the lower limit of 68 °C, while maintaining the relative humidity at 95%, condensation would indeed occur in the chamber by the time the temperature dropped near 70 °C, the specified temperature (as indicated by the tolerance bar crossing the saturated vapor curve to transition from vapor to liquid water). Reexamination of possible conditions lead to the decision to reduce both the temperature and humidity specified in the protocol. The temperature was reduced to 65 °C and the relative humidity was reduced to 80%. These conditions were selected based on the vehicle data study which had indicated that temperatures of 65 °C were not unreasonable with respect to the potential operating environment of the armor, and the previous work performed at 65 °C. The relative humidity was reduced to 80% in an effort to avoid the issues with condensation. Figure 3.17 shows the tolerance associated with a drop from the upper limit of 67 °C to the lower limit of 63 °C, while maintaing the relative humidity at 85%. In this case, the tolerance does not touch the saturated vapor curve until the temperature drops very near 63 °C, so condensation is much less likely with these revised conditions. These changes were included in the soft armor conditioning protocol released in NIJ Standard–0101.06 in July 2008.

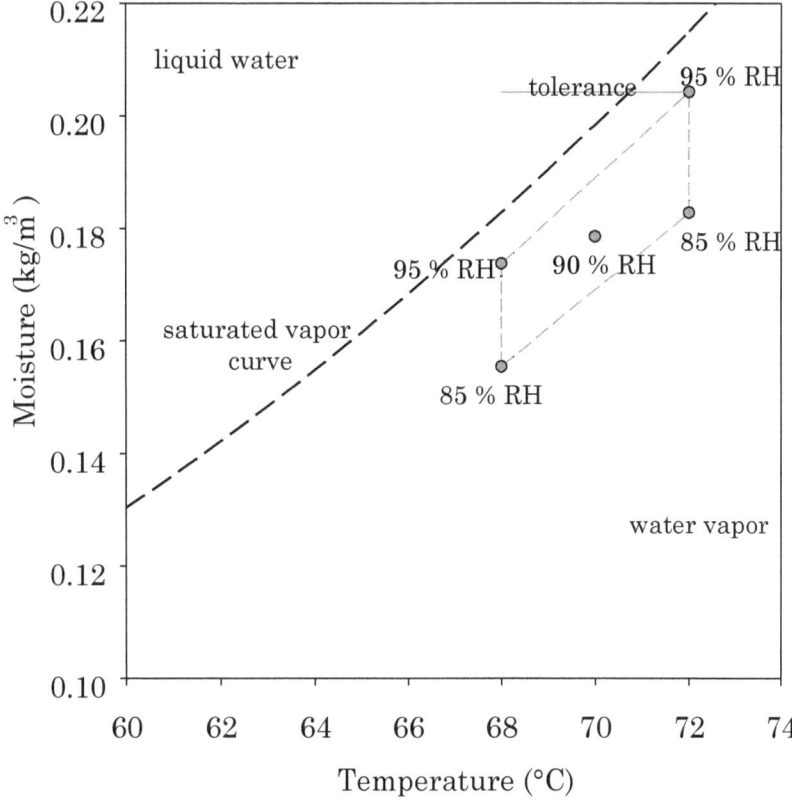

Figure 3.16: Analysis of Potential for Condensation at 70 °C and 90 % Relative Humidity.

3.9 Phase IV Summary

Minor adjustments to the protocol occurred in Phase IV. The environmental conditions were adjusted to prevent problems with condensation during minor, allowable excursions in conditions of temperature and relative humidity. The protocol used during this phase was adopted as Section 5, the Flexible Armor Conditioning Protocol in NIJ Standard–0101.06.

Development of Soft Armor Conditioning Protocols for NIJ Standard–0101.06: Analytical Results

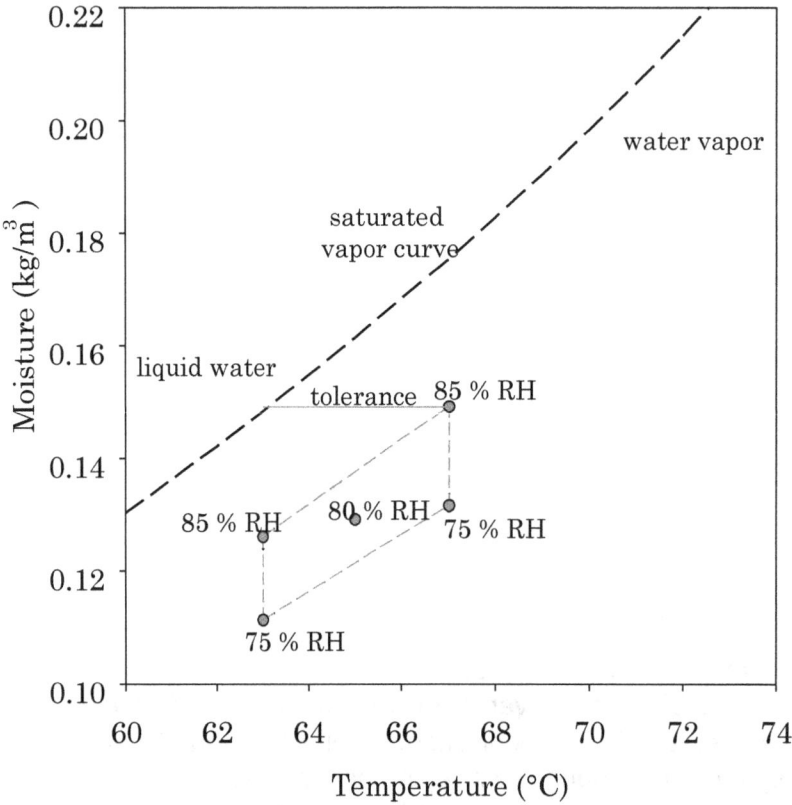

Figure 3.17: Analysis of Potential for Condensation at 65 °C and 80% Relative Humidity.

4 Conclusions and Future Work

After four major phases of development work, the Flexible Armor Conditioning Protocol in NIJ Standard–0101.06 (as it appears at the time of this publication) was finalized. The duration of the protocol was shortened from 9 weeks in the first efforts to 10 d. All major classes of materials were tested, and the conditions selected are found to be quite detrimental to PBO armors of designs that previously had exhibited problems in the field. The conditions have not been found to be excessively detrimental to other commonly used types of armor. However, the protocol does not represent an exact period of time in the field.

Current and future work will attempt to validate this protocol to reflect a period of field service for armor. Two major efforts will work to create this validation. The first of these is an extensive study to examine the aging properties of most of the ballistic fibers used in body armor, which will attempt to determine the correlation between natural aging and accelerated (elevated temperature and relative humidity) aging. Methods will be explored to better correlate artificial mechanical damage to actual wear. Additionally, a large study to examine the physical and ballistic properties of fielded armor is currently being planned and initial work on this project is underway. As opportunities arise, more studies to examine fielded armor in support of the validation of this protocol will be performed.

This page intentionally left blank.

5 References

[1] NIJ. Status report to the Attorney General of body armor safety initiative testing and activities. NIJ Special Report, 2004.

[2] NIJ. Supplement I: Status report to the Attorney General of body armor safety initiative testing and activities. NIJ Special Report, 2004.

[3] NIJ. Third status report to the Attorney General on body armor safety initative testing and activities. NIJ Special Report, 2005.

[4] J. Chin, E. Byrd, A.L. Forster, X. Gu, T. Nguyen, S. Scierka, L. Sung, P. Stutzman, J. Sieber, and K. Rice. Chemical and Physical Characterization of Poly(p-phenylene-2,6-benzobisoxazole) Fibers Used in Body Armor. NISTIR 7237, 2006.

[5] J. Chin, E. Byrd, C. Clerici, M. Oudina, L. Sung, A.L. Forster, and K. Rice. Chemical and Physical Characterization of Poly(p-phenylene-2,6-benzobisoxazole) Fibers Used in Body Armor: Temperature and Humidity Aging. NISTIR 7373, 2007.

[6] A.L. Forster, J.W. Chin, and M. Gundlach. Effect of bending and mechanical damage on the physical properties of poly(p-phenylene-2,6-benzobisoxazole)(PBO) fiber. Abstracts of Papers of the American Chemical Society, 231:274–POLY, 2006.

[7] G.A. Holmes, J-H Kim, D.L. Ho, and W.G. McDonough. The Role of Folding in the Degradation of Ballistic Fibers. Polymer Composites, in press, 2009.

[8] T.E. Bachner. Rational Replacement Policy – A Recommendation to the Law Enforcement Community, October 1985.

[9] D.E. Frank. Ballistic Tests of Used Soft Body Armor. NBSIR 86-3444, 1986.

[10] National Institute of Justice. Response to DuPont's Rational Replacement Policy, 1985.

[11] DuPont Kevlar Survivor's Club. Personal body armor facts book. Technical report, 1989.

[12] Police body armor standards and testing Volume I: Report. Congress of the United States Office of Technology Assessment, 1992.

[13] Police body armor standards and testing Volume II: Appendixes. Congress of the United States Office of Technology Assessment, 1992.

[14] E. Hoffman. Soft Body Armor. Police Product News, 1986.

[15] Armor Holdings. Interim Report on Body Armor Testing: Investigations and Actions. Technical report, 2004.

[16] DHB Armor Group. DHB Armor Group Armor Safety Report. Technical report, 2004.

[17] Z.N. Frund. Response to Request for Information: Artificial Aging of Armor, November 2005.

[18] H.S. Fogler. Elements of Chemical Reaction Engineering. Prentice Hall International Series in the Physical and Chemical Engineering Sciences. Prentice Hall, Englewood Cliffs, New Jersey, 2nd edition, 1992.

[19] F.J. Stadler, J. Kaschta, and H. Munstedt. Dynamic-mechanical behavior of polyethylenes and ethene-/alpha-olefin-co-polymers. Part I. alpha'-relaxation. Polymer, 46:10311–10320, 2005.

[20] S. Chabba, M. van Es, E.J. van Klinken, M.J. Jongedijk, D. Vanek, P. Gijsman, and A.C.L.M. van der Waals. Accelerated aging study of ultra high molecular weight polyethylene yarn and unidirectional composites for ballistic applications. Journal of Materials Science, 42:2891–2893, 2007.

[21] Rheometrics Series User Manual. RSA III Manufacturer's Specifications. TA Instruments, New Castle, DE, pn 902-30027 edition, 2003.

[22] Ballistic resistance of personal body armor. NIJ Standard–0101.06, 2008.

[23] Ballistic resistance of personal body armor. NIJ Standard–0101.04, 2000.

[24] Ballistic resistance of personal body armor. NIJ Standard–0101.04 Interim Requirements, 2005.

[25] M. Riley. Environmental Conditioning Tumbler Reference Design, ftp://ftp.nist.gov/pub/bfrl/riley/NIJ -010106/Tumbler/, 2009.

www.ingramcontent.com/pod-product-compliance
Lightning Source LLC
Chambersburg PA
CBHW081738170526
45167CB00009B/3871